KU-229-693

A BUYER'S GUIDE TO

OLIVE OIL

Anne Dolamore

GRUB STREET • LONDON

ACKNOWLEDGEMENTS

I wish to thank all of the shops and shippers who so generously supplied me with the samples of olive oil to taste for this book, as well as much useful information. Also The International Olive Oil Council, The Italian Trade Centre and Foods From Spain. Special thanks to David and Jan Gordon in New York; Mark Lewis of Harvey Nichols; Charles Carey, The Oil Merchant; Alice Seferiades at Odysea; Bob Ferrand of Good Food Retailing and Tim Imrie for photographing the oils to such good effect. Glyn Christian's *New Delicatessen Food Handbook*, *Degusto* by Giuseppe Grappolini and *Start to Taste* by Max Lake were also useful sources of information.

FOR JOHN AND AMY

Published by Grub Street
The Basement, 10 Chivalry Road, London SW11 1HT

Photographs: Tim Imrie
Design and jacket design: Nicci Walker

British Library Cataloguing in Publication Data:
Dolamore, Anne
Buyer's Guide to Olive Oil
I. Title
641.3

ISBN 0-948817-80-1

Printed and bound in Slovenia by Gorenjski tisk

CONTENTS

FOREWORD

It is six years since I wrote *The Essential Olive Oil Companion* and in the intervening time I have marvelled at how the choice of olive oils has grown - in fact sales in the UK have doubled in the last five years. There could be a number of reasons for this huge increase but I think there are two especially. Firstly the general concern for healthier eating which has focused attention on the much acclaimed benefits of a Mediterranean diet, which is of course generous in its use of olive oil. There has also been, in certain quarters, fascination with a subject which has allowed a snobbery and a whole new vocabulary to develop around it. Happily these two elements of health and fashion have combined justifiably to raise the profile of a wonderful product, with the consequence that the range of oils now available to shoppers is large but possibly daunting and confusing.

When I researched my first book on this subject supermarkets had one, maybe two different olive oils and the only places providing a sizeable selection were delicatessens, and even then they were almost exclusively Italian oils. What a change there has been! Now supermarkets have own-brands of not just any old extra virgin olive oil but their own-brand Spanish, Greek and Italian varieties. And the food halls of some large department stores carry as many as fifty different extra virgin olive oils.

So the choice has grown and much has been written about the joys and wonders of olive oil but I still find friends and acquaintances bemused about the basics of what to buy and what exactly they should use a bottle of extra virgin for. The question I'm always asked when I give talks or run tutored tastings is - which is the best oil?

Herein then lies the problem - people are uncertain about what they might be buying. They think, "My god I'm not paying the equivalent price of a very expensive bottle of claret, for half a litre of olive oil only to find I don't like it." Quite right. We have now all grown so accustomed to the helpful comments on taste and serving suggestions we get in every high street wine shop, then why not for olive oil?

It has become obvious to me that people are eager to learn but they lack the guidelines. So I have written this book as a direct result of the very clear need to provide some guidance to the consumer as well as of course the fine food store owner who may equally be baffled about what to stock.

I have sniffed and tasted well over 100 extra virgin olive oils in compiling this guide; it has helped me to decide what I like and I hope therefore it will help you too. However, taste is notoriously subjective and you may not always agree with my opinions. I have to confess to a preference for the more assertive rather than subtle style of olive oil, but I have tried to be as objective as I can about the quality of an oil even if I didn't like its particular style because there are well made and badly made oils. I hope that my findings will encourage you to try some of my recommendations and thereby discover for yourself what your favourites are. If you disagree with me that's fine because ultimately the best oil is the the oil *you* like best.

....

INTRODUCTION

In my previous book I related the fascinating and ancient history of olive oil, when it first appeared and how its story is woven amongst the myths and legends as well as the everyday life of the Mediterranean. I also included many of the recipes from each olive growing country.

My aim in this book is to provide a practical guide to what olive oil is and how it is made, so that when you go to buy olive oil you will know how to make informed choices and having bought your oil, to use it to best advantage.

I always have at least three different olive oils in use in my kitchen. There will be several bottles of the very best first cold pressed single estate oils for their fabulous variety of flavour and aroma. These I use to drizzle raw over cooked dishes such as soups, casseroles, pasta, cooked vegetables, especially potatoes, or salads; a bottle of commercially blended extra virgin at the cheaper end of the price range for basting fish or meat when cooking, for marinades or for making highly flavoured salad dressings say with lots of herbs, spices or garlic and then finally a 5 litre can of olive oil (which is refined olive oil and used to be called under old EC regulations, Pure Olive Oil). This olive oil, even though it is refined, is not to be despised. People have got so hung up these days on extra virgin olive oil they seem to ignore this grade but it has an important culinary role, for frying or making mayonnaise. I never use extra virgin for making mayonnaise, I think it makes a too rich, sickly mixture. I suggest making the emulsion with olive oil and then, if you want more flavour or piquancy add drops of extra virgin to reach the required taste.

So that's my store cupboard of olive oils and I would advise that if you want a basic start-up kit, so to speak, you arm yourself with the same; a single estate first cold pressed, a commercial blend and olive oil. But what do each of these terms really mean and what have you got in the bottle? And why should you use olive oil rather than the other vegetable or seed oils available?

OLIVE OIL AND HEALTH

Oils and fats are essential in a well balanced diet as they are a rich source of energy. But which type of fat we consume has been shown to have a direct bearing on the incidence of cardiovascular illnesses. Recent research has provided incontrovertible evidence that a

Mediterranean-style diet rich in olive oil is not only generally healthy but that consuming olive oil can actually reduce cholesterol levels. Low fat diets have long been recommended to reduce cholesterol, with the emphasis on cutting back on animal fats (saturated fats) and using olive oil and other vegetable fats (unsaturated fats).

OIL AND FAT COMPOSITIONS

	% Saturated	% Monosaturated	% Polyunsaturated
Coconut oil	92	6	2
Olive oil	12	80	8
Corn oil	16	27	57
Sunflower oil	10	18	72
Safflower oil	12	10	78
Butter	58	39	3
Margarine	64	30	6

For many years it was polyunsaturates which were felt to be best for you but in 1986 the results of research into monounsaturated fatty acids (in which olive oil is especially high) came up with amazing evidence about the nature of cholesterol. There are two types, low density (LDL) and high density (HDL). Low density lipoproteins (LDL) transport and deposit cholesterol in the tissues and arteries. LDL increases with excessive consumption of saturated fatty acids and is therefore potentially harmful because it will deposit more cholesterol. HDL on the other hand eliminates cholesterol from the cells and carries it to the liver where it is passed out through the bile ducts. Polyunsaturates reduce both LDL and HDL but monounsaturates reduce LDL while increasing HDL. An increase in the level of HDL will not only provide protection against cholesterol deposits but will actually reduce cholesterol levels in the body.

Besides being a monounsaturated fat, virgin olive oil, because it is a completely natural untreated food, is rich in antioxidants and vitamins, which help prevent body cell ageing, as well as giving the oil itself its conservation properties.

COMPOSITION OF OLIVE OIL

Vitamin E (3-30mg)
Provitamin A (carotene)
Monounsaturated fatty acids (oleic) 56-83%
Polyunsaturated unfatty acids (linoleic) 3.5-20%
Polyunsaturated fatty acids (linolenic) 0-1.5%
Saturated fatty acids 8-23.5%
9 calories per gram

The body can produce most of the fatty acids found in foods but it cannot make either oleic acid or linolenic acid. These are termed essential fatty acids and as you can see from the table above olive oil contains two of these making it an important and valuable dietary ingredient.

Since olive oil also contains Vitamin E (one of the currently fashionable vitamins) and oleic acid which is also found in maternal milk, it aids normal bone growth and is particularly suitable for expectant and nursing mothers because it encourages the development of the infant's brain and nervous system before and after birth.

From all this evidence doctors and nutritionists recognize extra virgin olive oil as having the most balanced composition of fatty acids (saturated, monounsaturated and polyunsaturated) of all edible vegetable oils.

FIRST COLD PRESSED EXTRA VIRGIN

These extra virgin oils, very often from a single estate, are the best olive oils you can buy and as a consequence they are at the top end of the price range. They can cost anything from £10 -£30 per litre. The reason for the high price is the quality you are getting and the great care that has gone into producing them, often in very limited quantities. They are not produced in bulk and are usually made by the people who have grown the olives and had them pressed in their own mill. They are the equivalent of top growth wine or finest champagne, and you should never, never cook with them. It's a waste. They should be used as a condiment, as a heavenly flavouring poured over salads or dishes when they are cooked, so that the heat of the dish brings out the gorgeous flavours. I almost never mix them with vinegar or lemon juice, because to my mind their flavours are so exquisite on their own they need nothing more. After all you wouldn't think of mixing a premier cru claret with lemonade would you? If you are amazed or appalled by the idea of using finest olive oil on its own then you haven't yet explored this essential ingredient.

COMMERCIAL BLEND EXTRA VIRGIN

An important distinction to be aware of with extra virgin olive oils is what we term commercially blended extra virgin. These extra virgins can be likened to vin de table. They are the brands you will see in almost every supermarket and independent grocer such as Berio, Sasso, Napolina, Menucci, Bertolli, Cypressa and Carbonell, as well as, of course, the supermarket's own-brand extra virgin oils. These companies buy in olive oil by the tanker load from local producers and cooperatives and blend them. They will, as a consequence, taste the same year in year out because, just like non-vintage champagne, they are blended to a particular style. These oils are considerably cheaper than first cold pressed oils because they are produced in bulk, the average price being £4-5 per litre. Some of them are of excellent quality and thereby exceptional value for money, but sadly most of the supermarkets seem to stock their shelves with the same ubiquitous Italian brands whose quality often leaves much to be desired. I would really make a plea to those responsible for supermarket buying to be a bit more adventurous in their purchasing decisions, as they have been of late with the standard of their own-brands, for which I commend them. They really must search out the best commercial extra virgin olive oils for their customers because the supermarket is where the great majority of the population make their olive oil purchases and I think people have the right to experience the best that is available in this price band. At the moment some of the best value-for-money extra virgin olive oils in this category come from Spain and Greece.

These commercial blends of extra virgin are everyday oils. Use them to make salad dressings with lemon or vinegar or other flavourings. They are great for marinades with herbs and spices or basting food while it is cooking, such as chicken or grilled fish.

Both first cold pressed extra virgin olive oil and the commercial olive oils are quite simply the juice of crushed, pressed olives and their flavour and aroma, like grapes depends on:

- variety
- soil
- climate
- cultivation
- health of the fruit
- degree of ripeness
- method of harvesting
- extraction method
- country of origin

Oils will vary in taste and colour from year to year, and even from one pressing to another. There is no such thing as a vintage in the wine sense with olive oils but of course some years will be better than others and it may well be that if you develop a liking for a particular first pressing extra virgin oil you may find a year that it is not to your liking.

This is important to remember because my tasting notes for the first cold pressed extra virgins refer to the oils from the current harvest and it may well be that next year's harvest will give a very different oil. However, I would expect the style to be similar because that is dictated by the olive varieties.

OLIVE VARIETIES

There are so many different varieties of olives, varying according to the country or even region and like grapes they have different qualities and flavours. Some of the more common ones used for making olive oil are: Arbequina and Picual,in Spain; in Italy almost too many to mention but better known ones are Frantoio, Leccino, Moraiolo and Coratina; La Tanche and Picholine,in France; and Koroneiki,in Greece. These are the olive varieties which are prized for pressing. In addition there are equally as many olives which are best for eating such as Manzanilla or Kalamata.

You will notice more and more that top quality olive oils carry the names of the olives used and just like varietal wines you can now find single variety olive oils, especially amongst the Italian oils. You can compare just how much the soil or region will affect the taste if you try olive oils made with the same olive varieties but from different areas. Compare for example an oil from Tuscany such as Dell'Ugo made from Frantoio, Moraiolo and Leccino olives, which has a lovely,leafy bitter flavour and Tenuta di Saragano from Umbria made with the same olive varieties but with a flavour which is soft and warm and chocolatey.

HARVESTING AND EXTRACTING

So the olive variety will provide differences of taste and fragrance in your bottle of extra virgin but what else?

Well, the method of harvesting and pressing will certainly have an impact on the quality of the resulting olive oil. The best olive oils are made from hand picked olives at the optimum level of ripeness which usually means while they are still green and certainly before they start falling from the trees. If olives are shaken from a tree and fall to the ground they bruise and bruised fruit will start to oxidise and ferment, raising the acidity level of the resulting oil and as you will discover later on, acidity levels are the criteria by which olive oils are categorised.

Each variety has its peak of maturity, at which stage the olives need to be harvested in order to extract the best oil. Olives which are too green give an intensely bitter taste and those which are too ripe or black produce oil with a flat taste and often a high acidity.

If you've ever wondered at the price of the finest extra virgin olive oils then the fact that they are hand picked should reveal all. Imagine the labour involved and thereby the cost of going over an entire olive tree and not just one olive tree but thousands upon thousands of olive trees and you'll have some conception of why you'll pay dearly for the best olive oil. But it will certainly be worth every single pound and just think if you're prepared to spend £6 for a bottle of very good wine, maybe even £30 for an excellent bottle of wine which will last you just one meal or one evening, how much more value there is in a bottle of olive oil at a comparable price when it will last you months. Unlike wine, however, it will not improve with age, so don't bother laying it down - it should really be used within the year or by the 'best by' date.

Harvest time, depending on the olives and the country, is between September and December but may go on to January or even February. The olives are finally taken to the mill where they may be stored for a while to allow them to heat up a little, helping finally to release the oil from the crushed fruit, but certainly for no longer than a day or so, otherwise they will start to ferment, which would ruin the resulting oil. The olives are then washed to remove leaves, twigs or earth, and crushed to produce an homogenous mixture from which the liquid can be extracted. The paste obtained by crushing the olives is kneaded mechanically to help the amalgamation of the minute droplets of oil found in the pulp. This resulting mixture is a combination of liquid (oil and water) and solids (pulp and stones). Between 96 and 98 per cent of the oil is contained in the flesh, with only 2 to 4 per cent in the stone. Roughly 15 to 25 per cent of the fruit is oil, whereas the quantity of water varies between 30 to 60 per cent. Of the rest 19 per cent is sugar, 5 per cent fibre, 1 per cent protein and the remainder various mineral elements.

There are two basic methods of extraction. The first is called **traditional** and involves the extraction of the oil by mechanical pressure. The pulp of the olives is spread in thin layers, separated by round filters made of synthetic material. These are then placed in an hydraulic press with pressure varying between 250 and 400 kg per square centimetre. The oil runs off from the solid matter, filters through and drips into containers, where the oil and water are separated most often these days by means of a centrifuge.

The second method is called **continuous** where extraction is entirely by centrifuge. Here the paste is spun at high speed to separate the flesh and the oil. Both methods are equally good, though many people ntinuous extraction gives a greater consistency of quality.

When the oil has been separated from the water you are left with first cold pressed virgin olive oil, which is a totally pure product because it is untreated. Cold pressed means that the temperature during the oil extraction process has been controlled not to exceed 30°C. If the temperature rises above this level the quality of the oil will be affected. No other vegetable oil is edible just by being pressed. All other oils have to be treated first because they contain toxins or are not suitable for human consumption in their natural state.

You may also be interested to know that it takes 5 kilos of olives to produce 1 litre of oil and each tree can produce between 3 to 4 litres of oil.

COUNTRY BY COUNTRY GUIDE

Every country around the shores of the Mediterranean produces olive oil, as well as Australia and California and the style of the oils is different in each one. But here I shall just concentrate on the countries whose oils most often find their way into our shops. Generally speaking the oils of France are delicate and subtle, sweet and gentle; in Greece they are grassy and green; in Italy they range from the strongly assertive and peppery in Tuscany up in the north to the fruitier ones of Apulia right down in the south; in Portugal rustic and earthy and in Spain bursting with the flavours of tropical fruits or almond. Though there are within each of the countries strong regional differences also.

FRANCE

You may be surprised to discover that France is actually one of the world's smallest producers of olive oil. Consequently, while more French oil is now finding its way into the shops it is not widely available but it has always been held in high regard. The provinces in the south and especially those bordering the Mediterranean make up the twelve departments where olives grow with the best areas considered to be in Provence. Throughout the region you will find hundreds of small olive mills serving the thousands of tiny producers. Nyons is the olive town of Provence, nestling in the foothills of the Alps on the banks of the River Aygues, where the local variety, Tanche, thrives in the microclimate here. The oils are sweet and fruity and Huile d'olives Nyons has its own Appellation d'Origen.

Travelling south towards the coast you come to Maussane beneath the hills of Les Alpilles. This has a famous co-operative, thought by many to produce the best olive oil in the whole of France,with a strong, distinctive flavour sold under the name of La Vallee des Baux and made predominantly from the Picholine variety. One other olive oil of international repute is from Monsieur Alziari and sold from his shop in Nice. Made from olives grown locally it has a sweet and delicate aroma and a flavour to match. Generally if you come across an oil marked Huile de Provence you're pretty well assured of quality. I don't always like the style of some of the French oils as you will see from individual tasting notes but the good ones are so different and distinctive and deliciously sweet that I strongly recommend you try them if you've only ever sampled Italian oils.

GREECE

Greece is the third largest producer of olive oil in the world and the Greeks consume more olive oil per head than any other country. Which only goes to show how wise they are because I cannot recommend too highly the extra virgin olive oils which are currently on sale here. They are generally of remarkable quality at truly amazing prices. For many years the Greek oils did not find favour with our palate; they were described as rustic and assertive, though personally I always liked them. However the oils coming into the country now have been blended to a lighter, less aggressive style and clearly the Greeks have worked hard at establishing real quality. The main producing areas are the Peloponnese and Crete, where the most important olive variety for oil making is Koroneiki. If you're thinking I've forgotten Kalamata, this is the town in the Peloponnese which gives its name to the best variety for eating.

ITALY

Without question some of the most sublime single estate bottled oils come from Italy and not surprisingly since Italy has more varieties of olives than anywhere else in the world. Every region produces olive oil, though the world at large may still know only of Tuscany and its celebrated town of Lucca. Many of the Tuscan oils are indeed gorgeous, reminding me often in aroma and taste of fresh, green meadows. Up front and assertive with greater or lesser degrees of peppery aftertaste, sometimes the pepper is too much for me and I find myself coughing from its powerful kick. Oils from this region are inevitably the products in the main of Chianti growers and you will find the gallo nero logo reproduced sometimes on their bottles of oil. One of the latest marketing innovations from the region is the Laudemio consortium - a group of Tuscan producers who sell their own individual oil in the same over-sized scent bottles - but overall I was not impressed by the quality of the oils. I couldn't get away from the feeling that I was being offered high priced packaging. My advice with Italy is be adventurous, branch out and try, just for a change, the oils I have recommended from some of the other regions such as Umbria, which grows many of the same olive varieties found in Tuscany but which are fruitier and less green and bitter; Liguria, whose oils are light and fruity, and Apulia, the country's largest producer mostly from Coratina olives. The oils are delicious and full of the flavour of ripe olives.

PORTUGAL

I tasted only two oils from Portugal in the course of writing this book, and sadly I did not like the style of either. They are indeed earthy and rustic but I felt unpleasantly so. My opinon is there is some work to be done on the style and quality of these oils.

SPAIN

How gorgeous are the best olive oils from Spain. Oils from the north often have a lovely bitter almond taste while in the south they are packed full of luscious tropical fruit scents and flavours. What is remarkable about Spanish oils is that despite being the world's largest producer there seems to be no sacrificing quality. In fact the Spanish were the first to establish demarcated regions of production, known as Denominacion de Origen, for their olive oils thereby guaranteeing quality. There are four DO areas: Borjas Blancas and Siurana in Catalonia, where the Arbequina olive tends to prevail, and Sierra de Segura and Baena in Andalucia, where the oils are more commonly made with Picual olives. However it has to be said that the quality of olive oil from outside the DO areas is always remarkably good. I love the Spanish oils for their harmonious balance.

BUYING OLIVE OIL

When you come to choose a bottle of olive oil you will see labels bearing the words extra virgin, occasionally virgin and then olive oil. To really understand what each of these are it helps to know that virgin olive oil is graded according to its acidity in the following categories set down by the International Olive Oil Council.

VIRGIN OLIVE OIL is the oil obtained from the fruit of the olive tree solely by mechanical or other physical means under conditions and particularly thermal conditions, that do not lead to alterations in the oil. Further the oil has not undergone any treatment other than washing, centrifugation and filtration.
Virgin Olive Oil fit for consumption as it is, is classified into:
EXTRA VIRGIN OLIVE OIL is virgin olive oil of absolutely perfect taste and aroma and a maximum acidity in terms of oleic acid of 1%, though

many of the best are 0.5% acidity.

FINE VIRGIN OLIVE OIL is virgin olive oil of absolutely perfect taste and aroma having maximum acidity in terms of oleic acid of 1.5% or less.

ORDINARY VIRGIN OLIVE OIL is virgin olive oil of good taste and aroma having a maximum acidity in terms of oleic acid of 3%.

Virgin olive oil not fit for consumption as it is:

VIRGIN OLIVE OIL LAMPANTE (lamp oil) is an off tasting and or off smelling virgin olive oil with an acidity in terms of oleic acid of more than 3.3%. It is intended for refining and becomes one of the following:

REFINED OLIVE OIL is olive oil obtained from virgin olive oils by refining methods.

OLIVE OIL is a blend of refined olive oil and one of the top three grades of virgin olive oil to give flavour and aroma. It has an acidity in terms of oleic acid of no more than 1.5%. This oil under the old regulations used to be called Pure Olive.

So you can see from the above that the lower the acidity level the better the quality of the oil. The acidity of an oil is important also because it affects the speed at which the oil will deteriorate. The only necessary qualification however for a virgin olive oil to rate being called extra virgin is its acidity and here is the rub; it is possible to reduce acidity by chemical methods, which means lesser quality oils can be promoted into what should be the premier league. Some extra virgin olive oils may not be all that they should be. My message therefore, when buying extra virgin olive oil is *caveat emptor*.

The industry and the controlling bodies need to come up with a definition that really is a bench mark of quality which the consumer can trust but in an industry that now represents enormous profits especially to those in the top league of exporters, there would appear to be no incentive to set this particular house in order. In my experience of tasting all the extra virgin olive oils for this book there is rather too much indifferent quality commercially blended extra virgin oil on the market, with none of the fruity aroma and taste of ripe olives which should be found there. Amongst the oils I tasted it appeared to be just a few Italian producers who seem most guilty in this area. The Spanish and the Greek oils I sampled were consistently of remarkable quality at very low prices. My guess is that some of the Italian brands, because they have cornered the market for so many years by lots of advertising and fancy bottles, having lured the innocent consumer into thinking that the only decent oils are Italian, are just sitting back on their reputations.

One other twist to this complex question of precisely what you are getting in your bottle of extra virgin oil, is the present EC legislation which allows countries to import olive oil, bottle it and re-export it without having to declare the country of origin of the oil. Italy imports Spanish and Greek olive oils in enormous quantities, blends them with Italian olive oil and sells it abroad as produce of Italy. This is not a criticism and there is nothing underhand in this practise, it's just that you need to realise that just because it says Italian on the label it may not be that the oil has actually orginated in the country.

It is a commonly held myth that you can tell what an oil is like by its colour. For the record colour, which can range from yellow to intense green, gives absolutely no indication of taste or quality, which is dependent on olive variety and degree of ripeness. A green olive will give a different coloured oil to a ripe black olive and while I'm on olives you may also not know that there is no such thing as a green olive variety and a black olive variety as with black and white grapes; every olive starts green and when it's ripe it's black. So forget about colour because in some places they crush a few olive leaves with the olives to make the oil greener! The only hint you may glean about an oil from its appearance is that if the oil is opaque or hazy, there is a pretty good chance the oil hasn't been filtered, which is fine as it simply means there are bits of olive in the oil.

FLAVOUR AND HOW TO TASTE

So what about the wonderful flavours I have mentioned. There are flavours in these extra virgin oils you might never imagine could come from simple crushed olives - a taste of tropical fruit in some, such as melon, lychee or even banana, in others the deep tones of caramalised nuts, chocolate even, green grass, tomato, or apples.

To decide which olive oils suit your palate you need to try a number. So how do you go about this? Well, you can go to olive oil tastings, which are now regularly run by many of the better food shops and wine merchants. Or of course just go out and buy a small bottle, take it home and try it yourself. Interestingly some of the supermarkets like Safeway now sell tasting size bottles of their own-brand extra virgin olive oils at around 50p a bottle. You will see too from the list of shops, the places which give you an opportunity to taste before you buy.

The way to taste is simple. Just pour a little oil, say a few tablespoons, in a glass. Warm the glass, and thereby the olive oil, in your hands for a few moments to release the volatile aromas in the oil. Bring the glass to your nose and slowly inhale two or three times, taking in the fragrance. It may be complex or it may be simple but above all it should be fresh.

There are then two ways of tasting the oil. You can either dip a piece of plain white bread into the oil and taste it or my preferred method is to take a small sip of the oil from the glass, allow the oil to slide onto your tongue but do not swallow yet. After a few seconds form your tongue into a spoon shape and position it towards your top teeth, now with your mouth slightly open inhale two or three times in quick succession. The mixture of air and oil will spray your mouth and palate, allowing you to register the sensations of the flavours. Try to store in your mind the immediate impressions of sweetness or bitterness, merits or defects. It is then worth repeating the exercise having cleansed your mouth with fizzy water or a slice of apple, the acidity of which will cut through the oil.

There are four main taste experiences: sweet, acid, and salty which come from the top and sides of the tongue, and bitterness which is registered by the throat. As the whole palate registers different sensations, it is necessary to swallow the oil to complete your final impression of the oil which may be different to the first because there is also the aftertaste.

At any organised olive oil tasting there should be no more than six oils and this is how I tasted the oils for this book, in groups of six. It is advisable to set out each oil with some in a white saucer to examine the colour and also in a wine glass to enable you to sniff the aroma. You should also supply teaspoons for tasting from and abundant quantities of cubed white bread for dipping, sliced apples, paper napkins for wiping the teaspoons and fizzy water.

Having sniffed and tasted, how then do you describe what you have experienced? There is an official vocabulary drawn up by the International Olive Oil Council for their professional graders and tasters, whose job it is to assess the organoleptic qualities of olive oils and some of these words, which the ordinary taster can also utilise, I give below along with other terms I have coined through my own experience of numerous tastings.

Styles of olive oils vary from the sweet, through the ripely fruity to the assertively green and bitter, with degrees of pepperyness. But remember bitter is not a critical term with olive oil.

Styles of oils

Aggressive, Assertive or Pungent: oils which have strong up-front flavours or aromas
Bitter: characteristic taste of oils obtained from green olives or olives turning colour. It can be more or less pleasant depending on its intensity
Delicate or Gentle: a light combination of flavour and fragrance
Fresh: a sensation of freshly squeezed fruit with a significant aroma
Fruity: reminiscent of both the flavour and aroma of sound fresh fruit picked at its optimum stage of ripeness
Green: often used instead of bitter
Harmonious or Balanced: where the fragrance and taste are in perfect equilibrium
Rustic or Earthy: can be unpleasant or hearty and vegetal
Spicy: slight spiciness present in oils in the first months after crushing. Indication of healthy fruit
Strong: intensity of aroma or taste
Sweet: pleasant taste or aroma, not exactly sugary but found in oil where the bitter, astringent and pungent attributes do not predominate

Aromas and Flavours

Fruity
Apple, Banana, Lychee, Melon, Pear, Ripe Olive,Tomato

Verdant
Eucalyptus, Flowery, Grass, Green leaves, Hay, Herby, Leafy, Mint, Violets

Vegetal
Avocado, Earthy, Rustic

Nutty
Almond, Brazil Nut, Walnut

Chocolatey

Defects

Rancid: the flavour common to all oils and fats that have undergone a process of auto-oxidation caused by prolonged contact with the air. This is an unpleasant taste and cannot be corrected
Soapy, Fatty, Greasy: leaving an unpleasant tactile sensation in the

mouth and on swallowing. Often the result of olives which have been attacked by olive fly grubs

Earthy: characteristic flavour of oil obtained from olives which have been collected with earth or mud on them and not washed. This flavour may sometimes be accompanied by a musty-humid odour

Flat: olive oil whose organoleptic characteristics are very weak owing to the loss of their aromatic components

OLIVE OIL IN THE KITCHEN

Having bought your olive oil it is best to consume it by the 'Use by' date on the container. Store it away from light and heat because they cause oxidation. You can keep it in the fridge but I don't advise it because it will start to solidify at low temperatures, which won't harm it but you will need to allow first pressed oils to come back to room temperature before you use them, just in the way you let wines breathe. And likewise if you see bottles of oil in a shop with a white solid layer at the bottom, don't worry; there is nothing wrong with it, this will only be because it has just been taken from a cold store.

If you do buy an oil which doesn't have much flavour or you don't much like then don't despair remember you can always turn it to good use by putting some herbs or spices in the bottle, such as a few sprigs of fresh rosemary, some peeled cloves of garlic or a small handful of dried chillies. You can experiment with whatever you have available but if you do use fresh herbs take them out of the oil after a week or so or they'll go mouldy because they aren't sterilized.

Many people say they never deep fry with olive oil, feeling that somehow corn oil or sunflower oil are better but in fact the smoking point of olive oil is about 210°C, which is higher than corn oil (160°C) or sunflower oil (170°C), so you can use a refined olive oil. Finally, as I mentioned before but I think it's worth repeating don't cook with your very best extra virgin oils, use them raw, as a flavouring.

Here is a selection of some of my personal favourites from amongst all the extra virgin oils, some are estate bottled and others are commercial blends but I think they represent real quality and I hope you will have the opportunity to try some of them. In the guide to individual oils please note that prices are necessarily approximate, they vary from shop to shop.

Best supermarket own-brand: Safeway Greek

Best other own-brands: Harvey Nichols
Carluccio's Ligurian

Best Spanish: Nunez de Prado
Oro Magina
Siurana
Carbonell

Best Greek: Iliada
Solon

Best French: The Fresh Olive Company of Provence
Huile d'Olive Aux Anysetiers du Roy
Alziari

Best Italian: Callisto Francesconi
Rocchi Oro
Sasso
Santagata
Ravida
Monte Vertine
Tenuta di Saragano
Teja Cardinale
Santa Tea Fruttato Intenso
Santa Cristina
Taylor and Lake
Lungarotti
Fattorie Umbre Campomaggiore

Best value for money: Iliada
Oro Magina

HUILE D'OLIVE AUX ANYSETIERS DU ROY
France

Appearance Clear

Colour Rich golden

Aroma Apple, tropical fruit slightly floral with a fresh sharp quality

Taste Flat but not unpleasant, light style with a taste of Brazil nuts

Uses Salads, cooked vegetables

Price £7.50 / 500ml

Comments Bottle comes with yellow wax over the stopper which is a nuisance to remove since it gets everywhere. But it is packed in a dark bottle which is a plus. I like this oil, it's different and strikes me as unusual for a French oil. An interesting one to try. (FTC)

THE FRESH OLIVE COMPANY OF PROVENCE
France

Appearance	Clear and bright
Colour	Golden
Aroma	Strong and fruity and melon-like
Taste	Light and fruity. Hint of pepper on the finish
Uses	Salads, drizzled over vegetables, grilling, marinades
Price	£12.50 / 1 Litre
Comments	A lovely well made fruity French oil. Thoroughly recommend this. (F)

JEAN-MARIE CORNILLE DE LA VALLEE DES BAUX
France

Appearance Clear and bright

Colour Golden

Aroma Earthy, nutty, woody sweet

Taste Rustic and nutty

Uses Salads, grilled meats, marinades, grilled fish

Price £22.00 / 1 Litre

Comments I have to confess to not being over keen on this style of oil with its smell of wet rotting vegetation and earthy flavour but it's typical and well made if you like this. A bit on the pricey side though.(LL)

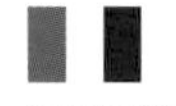

LOU NISSART
France

Appearance Clear and bright

Colour Deep golden

Aroma Faint sweet olive aroma

Taste No taste at all, very bland and slightly greasy on the palate

Uses Salad dressings with strong flavourings

Price £9.00 /1 Litre

Comments Can't really see any point in buying this. Doesn't have any qualities to commend it to me and there are better oils in the price range.

A L'OLIVIER
France

Appearance Clear and bright

Colour Golden

Aroma Sweet and fruity

Taste Subtle, light with a hint of greenness

Uses Salad dressings, for coating while grilling meat or fish or vegetables

Price £5.10 / 500ml; £9.00 / 1 Litre

Comments Typical of the style of French oils. This is very pleasant, the flavour just insinuates itself. A good everyday choice. (OM)

HENRI BELLON
France

Appearance Clear

Colour Rich golden green

Aroma Sharp hint of earthy eucalyptus

Taste Difficult to discern any true distinct flavour - not unpleasant but nothing to mark it out as French

Uses Salad dressings

Price £13.45 / 500ml

Comments Disappointing. I can't really see any reason for buying this oil especially at this price.

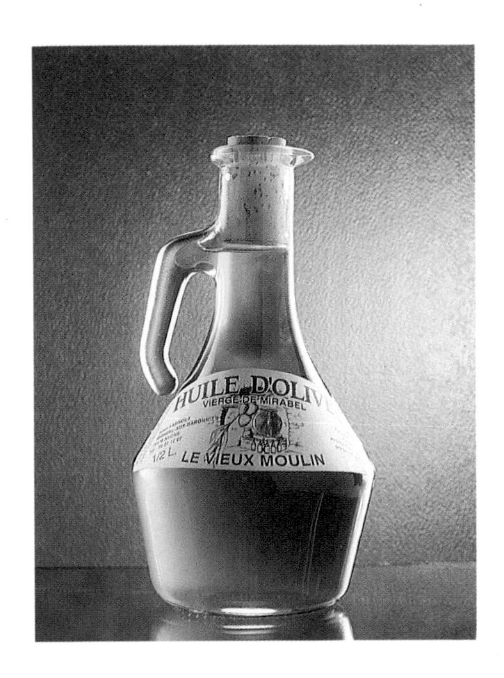

LE VIEUX MOULIN
France

Appearance Clear and bright

Colour Deep gold

Aroma Soft, gentle, more reminiscent of a nut oil. Maybe hint of almond or walnut but pleasantly earthy

Taste Greasy sensation, very little flavour

Uses Pouring over meat or fish for grilling. Salad dressings

Price £16.20 / 1 Litre

Comments I didn't like this oil, which comes from south of Nyons,it's too greasy on the palate. There are much tastier French oils around. (OM)

ALZIARI
France

Appearance Clear and bright

Colour Golden

Aroma Subtle, sweet, more reminiscent of a nut oil, maybe walnuts

Taste Sweet, light and fruity, maybe just a touch of ripe banana

Uses Drizzled over soups, casseroles, toss with vegetables, over salad leaves

Price £15.45 / 1 Litre tin

Comments A quite unusual flavour but one I really enjoyed. It is very different in style to Italian oils and it is certainly worth trying if you like your oils sweet rather than bitter. Comes fom Nice and is packed in a most spectacular tin which would make a gorgeous gift. (OM)

TAYLOR AND LAKE HUILE D'OLIVE VIERGE EXTRA
France

Appearance Clear and bright

Colour Dark golden

Aroma Sweet and fruity with green grass tones

Taste Subtle flavour, very light, pleasant

Uses Salads, soups, grilled vegetables or fish

Price £8-£9.00 / 500ml

Comments This is obviously a very well made oil from Vaucluse in Provence. It is well balanced, there is a subtle flavour but then this is the style of the French oils. If you want to use olive oil but don't like strong bitter or pronounced flavours they are probably for you. (TL)

E MSICA
France

Appearance Clear and bright

Colour Golden

Aroma Fruity, sweet touch of tropical fruit

Taste Light, subtle, quite pronounced pepper finish, sweet fruity

Uses Salad dressings, grilling fish, white meat

Price £3.95 / 1 Litre

Comments A good straightforward example of a light olive oil. As the label states packed in Provence and imported Product of EEC, I'd suggest it isn't entirely French oil; with the peppery finish, it may be a blend with Italian. Good value.

ILIADA
Greece

Appearance Slightly hazy

Colour Dark gold with a green tinge

Aroma Sweet and grassy with the scent of olive

Taste Light, mild taste of ripe olives

Uses It is ideal for anything where you don't want the flavour of the oil to dominate, just to insinuate

Price £5.00 / 1 Litre

Comments This is an excellent everyday oil and without a doubt my favourite Greek oil. Exceptional value for money. Highly recommended. (O)

KYDONIA
Greece

Appearance Clear and bright

Colour Rich golden

Aroma Faint hint of green grass

Taste Up front but disappears quite quickly, a little greasy on the palate, slight hint of pepper

Uses Salad dressings, grilling or roasting meat

Price £3.65 / 500ml

Comments There are many excellent Greek oils on the market now and I don't think this one, which comes from Crete, matches some of the others around. Rather disappointing because of its lack of flavour.

KOLYMVARI
Greece

Appearance Clear and bright

Colour Deep gold

Aroma Fresh green grass

Taste Pleasant, light, bitter olive

Uses Salad dressings, marinades, grilling meat, vegetables or fish

Price £3.50 / 1 Litre

Comments A good basic everyday Greek oil from Crete and amazing value for money. At this price you need never be without a good bottle of olive oil. This is cheaper than any supermarket brand I've come across. (B)

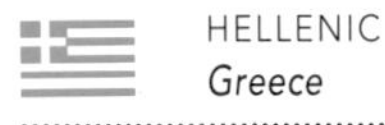

HELLENIC
Greece

Appearance Hazy

Colour Deep golden yellow

Aroma Bitter but faint

Taste Slightly greasy sensation at first but develops into a light, pleasant grassy flavour

Uses Making salad dressings, for grilling or roasting

Price £2.49 / 500ml; £4.40 / 1 Litre

Comments Greek oils are excellent basic everyday oils and real value for money. This one is a good example and generally available. (GS)

CHALICE
Greece

Appearance Slight haze

Colour Deep green gold

Aroma Sharp, fresh, lightly bitter with tones of green grass

Taste Very light and mild but a bit flat

Uses Salad dressings

Price £4.25 / 500ml

Comments Disappointing. It has a lovely aroma but the taste never really gets going for me and I'm left with a feeling that the blend is a bit too subtle.

SOLON
Greece

Appearance Bright and clear

Colour Golden green

Aroma Green leaf, green grass. Quite distinct

Taste Immediate pleasant grassy taste with light pepper after taste

Uses Salads, pouring over cooked vegetables, grilling fish

Price £4-£4.50 / 1 Litre

Comments Another fine example of a well made, pleasant, balanced Greek oil at a remarkable price. Excellent value. (JP)

MANI
Greece

Appearance Clear and bright

Colour Golden

Aroma Light green grass

Taste Light, subtle

Uses Salad dressings, grilling

Price £3.39 / 500ml

Comments Produced from olives grown on the Mani Peninsula. Doesn't taste of much at all, hardly worth buying. There is also an organic variety available which is fruitier and has more flavour.

CYPRESSA
Greece

Appearance Clear and bright

Colour Green gold

Aroma Lovely strong fresh green grass

Taste Subtle flavour of green grass, hint of pepper on the finish

Uses Salad dressings, grilled fish, raw sauces

Price £1.99 / 500ml; £3.59 / 1 Litre

Comments A very good example of an everyday simple, light style of Greek oil. (K)

ELEANTHOS
Greece

Appearance Clear and bright

Colour Green gold

Aroma Light green leaf

Taste Fruity green - not too assertive. A little greasy to the palate but not enough to be unpleasant

Uses Salad dressings, grilling, roasts, raw sauces

Price £1.99 / 500ml; £3.39 / 1 Litre

Comments Good, simple light style everyday oil. (K)

SAFEWAY EXTRA VIRGIN
Greece

Appearance Clear and bright

Colour Light green

Aroma Sharp, green leaves

Taste Delicate green leaves, first slightly greasy but not enough to be unpleasant

Uses Salad dressings, grilling meat or fish, or roasting

Price £2.49 / 500ml

Comments A good example of the new style, lighter Greek olive oils and excellent value for money. One of the best supermarket own-brands around.

WAITROSE EXTRA VIRGIN
Greece

Appearance Bright and clear

Colour Deep golden green

Aroma Sharp fresh green grass

Taste Quite pleasant typical Greek style, grassy and sharp

Uses Salad dressings, grilling meat or fish, roasting vegetables

Price £2.49 / 500ml

Comments This would be a good introduction to a not too robust Greek style of oil and is very good value for money.

SANTAGATA DEI COLLI D'ABRUZZO
Italy

Appearance Clear and bright

Colour Golden

Aroma Faint but fresh hints of grass

Taste Nutty deep tone, good olivey sweet hint of chocolate

Uses Salad dressings, marinades, poured over pasta, grilled meat or fish

Price £7.80 / 1 Litre

Comments Well made, simple, straightforward uncomplicated oil. Ideal for everyday use. Good value. (FS)

CARLUCCIO'S PUGLIAN
Italy

Appearance Clear and bright

Colour Golden green

Aroma Full fruity and green, lovely herbaceous nose

Taste Fruity medium weight oil with a subtle flavour. Quite strong peppery finish

Uses Salads

Price £16.50 / 750ml

Comments Nice oil, not an assertive style like Tuscan oils. Ideal for those who like a milder flavour. Packed in a wonderful stone jar. Great as a gift and great for protecting the oil from the light. (CO)

SANTAGATA DEI COLLI DI PUGLIA
Italy

Appearance	Clear and bright
Colour	Gold
Aroma	Good immediate fruity hint of banana and olive
Taste	Sweet, fruity and light
Uses	Salad dressings
Price	£7.80 / 1 Litre
Comments	Nice light easy everyday oil and as the name shows it comes from Apulia in the south. Not an overpowering taste, good weight and balance. (FC)

DEL CONSOLE CORATINA
Italy

Appearance Hazy - unfiltered

Colour Golden

Aroma Wonderful full green grass, reminiscent of freshly mown hay

Taste Sweet, round with a distinctive peppery finish

Uses Salads, poured over vegetables, soups, fish or casseroles

Price £9.00 / 1 Litre

Comments This is a lovely well balanced oil from Apulia and made from the Coratina variety of olive - not therefore as bitter as some Italian oils. Excellent value. (GFF)

RIVEROLI ORO
Italy

Appearance Clear and bright

Colour Gold

Aroma Light and green

Taste Light, subtle not much taste at all but a hint of pepper

Uses Cooked vegetables and fish, salad dressings

Price £3-£4.00 / 500ml

Comments Dark bottle. This oil, which comes from Apulia, is disappointing; it has very little taste, a touch too subtle. (A)

GAZIELLO MOSTO
Italy

Appearance Hazy - unfiltered oil

Colour Golden

Aroma Good strong green grass

Taste Lovely, subtle, fruity and green, hint of pepper on the finish

Uses Salads, grilled meat, fish or vegetables

Price £7.50 / 500ml

Comments A very good example of the different style of Italian oil, this is from Imperia. They are lighter, more subtle, less assertive than those from Tuscany. I like this very much. (F)

RAINERI ONEGLIA IMPERIA
Italy

Appearance Clear and bright

Colour Gold

Aroma Earthy, fruity

Taste Chocolatey, fruity, a touch greasy but not enough to be unpleasant since the flavours come through loud and clear

Uses Salads, drizzled over grilled vegetables, fish, meat

Price £5.70 / 500ml

Comments Packed in a dark bottle. A pretty good example of the dark chocolatey style of oil and good value. (FC)

ARDOINO FRUCTUS
Italy

Appearance Clear and bright

Colour Golden

Aroma Sweet and chocolatey

Taste Pleasant, medium weight and chocolatey

Uses Drizzled over vegetables, poured onto soups, casseroles, salad leaves

Price £10.27 / 750ml

Comments I like this oil though it may seem odd to describe an oil as having a chocolate flavour, as I have done also with other oils, but it's difficult to describe the sensation of a deep moody taste any other way. It really should be experienced. Plus points for the foil covered bottle as protection from the light. From Liguria.(GFF)

PRIMOLI OLIO LIGURE
Italy

Appearance Clear and bright

Colour Golden

Aroma Faint green leaf

Taste Sweet, light, simple and delicate with a hint of almond

Uses Poached fish, salads, vegetables, raw sauces, salad dressings

Price £13.50 / 750ml

Comments Good example of a simple straightforward subtle, delicate style of Ligurian oil made from Gentile olives. (FS)

ARDOINO VALL'AUREA
Italy

Appearance Hazy

Colour Golden

Aroma Faint green grass

Taste Subtle, light sweet fruity hint of pepper finish

Uses Raw sauces, salad dressings, over fish or vegetables

Price £10.90 / 500ml

Comments A very light oil from Liguria made from the single variety of Taggiasca olives but with little taste or aroma. I'm not very enthusiastic. (GFF)

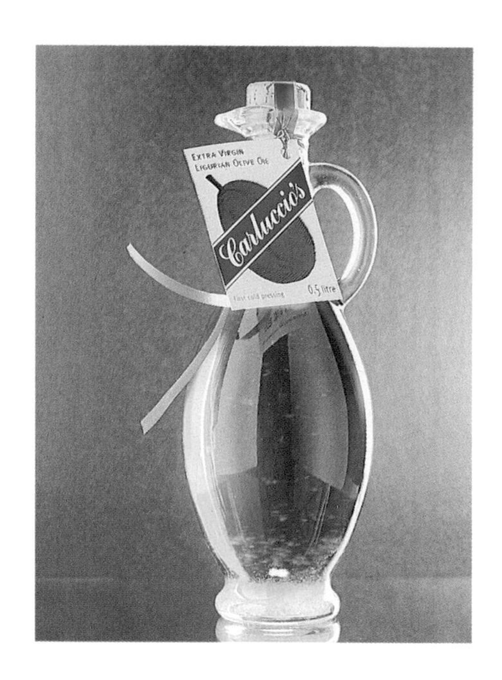

CARLUCCIO'S LIGURIAN
Italy

Appearance Clear and bright

Colour Golden

Aroma Light green grass

Taste Lovely light, subtle grassy flavour

Uses Where you want a subtle, hint of olive oil. Would be good in raw sauces, on salad leaves, steamed vegetables or fish

Price £7.00 / 500ml

Comments This oil seems to bounce off the tongue, doesn't cling to the palate at all. It's a lovely light-weight oil. (CO)

MARCO LE FASCE
Italy

Appearance Clear and bright

Colour Golden

Aroma Faint fruity olive

Taste Very light sweet fruity hint of pepper. Touch greasy to the palate

Uses Salad dressings, drizzled over fish or cooked vegetables

Price £7.00 / 500ml

Comments Nothing much to commend this rather ordinary oil from Liguria. (GFF)

MERCURI IL SAPORE DELLA OLIVE
Italy

Appearance Clear and bright

Colour Dark gold

Aroma Light aroma of green grass

Taste Almost imperceptible to start and then disappears to nothing

Uses Salad dressing and general cooking

Price £3.20 / 500ml amphora

Comments Nothing bad about it but nothing to commend it either even though it's quite cheap. From Marche.

MERCURI IL FRANTOIO
Italy

Appearance Clear and bright

Colour Green gold

Aroma Bitter, green olive

Taste Sweet and fruity olive flavours

Uses Salads, grilled vegetables, grilled fish, raw sauces, pasta

Price £5.50 / 500ml amphora

Comments A well made, medium weight oil. Fruity and sweet. From Marche (A)

VIOLA BARDOLINO BARDO
Italy

Appearance Clear and bright

Colour Golden

Aroma Faint green grass

Taste Pleasant, sharp, bitter with a very slight pepper finish - touch greasy on the palate, does not leave a distinctive flavour. Generally quite light.

Uses Salad dressings, grilling fish or meat, vegetables

Price £6.00 / 1 Litre

Comments A very light subtle oil but not very exciting. From Lombardy. (GFF)

PRIMOLI OLIO DEL GARDA
Italy

Appearance Clear and bright

Colour Golden

Aroma Good green grass

Taste Subtle green grassy, light, pleasant

Uses Drizzled over fish or vegetables or to dress salad leaves

Price £13.50 / 750ml

Comments Well made, well balanced and light weight. Lightly flavoured, subtle, delicate and gentle. This attractive oil comes from olives grown on the west side of Lake Garda. (FS)

PRIMOLI IL FRUTTO DELLA VITA
Italy

Appearance Clear and bright

Colour Greenish gold

Aroma Faintly fruity

Taste Fruity, deep tones of chocolate and nuts. Quite strong peppery finish

Uses Poured over meat, grilled vegetables, casseroles, soups, salads

Price £13.50 / 750ml

Comments This is an organic oil from Garda. I like its deep rich flavour - it's lovely. (FS)

BOTTARELLI LAGO DI GARDA
Italy

Appearance Clear and bright

Colour Golden

Aroma Light grassy

Taste Rather greasy, but generally a light oil with not much flavour, just a hint of grass

Uses Salad dressings

Price £12.50 / 500ml

Comments Not much to commend this. From Lombardy. (GFF)

COLONNA
Italy

Appearance Bright clear

Colour Green with gold

Aroma Subtle aroma of warm peppery olive and green banana

Taste Light buttery flavour. Hint of pepper on the finish

Uses Salads, fish, vegetables

Price £16.15 / 1 Litre

Comments Delicate, light, mild, harmonious estate bottled oil from Molise in Central Italy, made predominantly from Coratina olives. This is a lovely oil and good value for such a quality product. (OM)

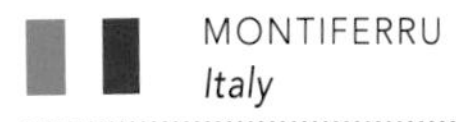

MONTIFERRU
Italy

Appearance Clear and bright

Colour Gold

Aroma Earthy

Taste Light, earthy with a hint of pepper on the finish

Uses Salads, poured over cooked meat or fish

Price £6.00 / 500ml; £8.50 / 750ml

Comments Reminds me of a French or Portuguese style oil, in fact it comes from North-West Sardinia. Not to my taste.

RAVIDA
Italy

Appearance Clear

Colour Golden with a hint of green

Aroma Wonderful tomato skin at the beginning mingled with bitter green grass moving to green apple. Lovely

Taste Delicious full bitter flavour of green grass leaving a lingering taste of green apples

Uses Use this on anything and everything but use it raw

Price Price £14 - £15.00 / 750ml

Comments A prize winning estate bottled oil from Sicily. Made from olives unique to the vicinity. It is a superb example of oil made to the highest standard and I would never tire of it. Worth every penny. (OM)

SAN REMO
Italy

Appearance Clear and bright

Colour Light, green

Aroma Sweet delicate with just a hint a green leaf edge

Taste Good light fruity flavour

Uses Drizzled over salads, or cooked vegetables, soups or fish

Price £8.00 / 750ml

Comments A gentle, well made, well balanced oil. (LL)

LAUDEMIO FRESCOBALDI
Italy

Appearance Clear and bright

Colour Lovely light green

Aroma Strong, sharp, fresh green leaves, hint of apple

Taste Light and green leafy with strong peppery finish

Uses Use raw drizzled over salads or cooked vegetables or poured over bread or pasta

Price £14.00 / 500ml

Comments Obviously a classy oil. I love the initial flavours but the peppery finish is a bit too much for my taste. From Tuscany. (OM)

BADIA ALBERETO
Italy

Appearance Thick and dense

Colour Deep, rich olive oil

Aroma Like falling face down in a field of new mown hay. Gorgeous, fresh, sharp

Taste Slightly greasy on the palate. Light and grassy

Uses Salads or drizzled over cooked dishes

Price £14.20 / 500ml

Comments After the heady, intoxicating aroma the taste is a little disappointing for a single estate bottled oil. I wonder about value at the price. From Tuscany. (OM)

RUFFINO
Italy

Appearance Clear and bright

Colour Golden green

Aroma Green, fruity, fresh and astringent

Taste Light, quickly gone, doesn't linger much. But it is chocolatey and nutty

Uses Salads, drizzled on grilled vegetables and fish

Price £14.75 / 750ml

Comments Good, well made harmonious oil from Tuscany. Light attractive style. (A)

ROCCHI OLIO EXTRA VERGINE DI OLIVA DI LUCCA
Italy

Appearance Clear and bright

Colour Golden

Aroma Green grass

Taste Light, subtle - a little greasy on the palate and the finish has a hint of pepper

Uses Salad dressings

Price £7-£8.00 / 750ml

Comments A disappointing oil. Its lightness and subtlety is overwhelmed by the greasy sensation. I much prefer the Rocchi Oro which is cheaper. From Tuscany. (A)

ANDREINI MAREMMA TOSCANA
Italy

Appearance Cloudy - unfiltered

Colour Green gold

Aroma Sharp, bitter green grass - delightfully fresh

Taste Smooth and mildly fruity, and bitter very faint peppery aftertaste

Uses Salads, drizzled over vegetables, fish, soups or hot dishes

Price £17.00 / 750ml

Comments A good oil, not very exciting especially at this price, but if you like milder oils this would be a good choice. Its plus points are that it is packaged in a dark bottle which helps protect against sunlight. (LL)

ANDREINI MIGNOLA
Italy

Appearance Clear and bright

Colour Deep rich gold

Aroma Apples and green grass

Taste Sharp pleasant, lightly fruity not very assertive, quite delicate

Uses Salads, poured over fish or vegetables

Price £20.00 / 750ml

Comments An expensive Tuscan oil but not exciting enough to inspire me. Packed in a dark bottle which is good but to my mind too pricey for what it is. (LL)

GIANCARLO GIANNINI OLIO TOSCANO
Italy

Appearance Clear and bright

Colour Golden

Aroma Fresh green grass

Taste Flat and faint, not much of anything

Uses Salad dressing mixed with strong flavours

Price £8.85 / 500ml

Comments Can't recommend this, it has no taste to speak of. (GFF)

SANTA CRISTINA
Italy

Appearance	Clear and bright
Colour	Golden
Aroma	Fresh green grass
Taste	Soft, subtle light fruit, hint of pepper finish
Uses	Salads, drizzled on fish, soups and vegetables
Price	£12.25 / 750ml
Comments	A very attractive harmonious oil on the gentle, lighter end of the scale. Quite lovely. Good value too. (EW)

MONTE VERTINE
Italy

Appearance Clear and bright

Colour Green gold

Aroma Gorgeous, full, immediate strong green grass

Taste Wonderful fruity green flavour with a good pepper finish

Uses Drizzled over salads, soups, hot dishes, grilled vegetables, bruschetta

Price £13.85 / 750ml

Comments Lovely estate bottled oil from the heart of Chianti. Well balanced, well made, just the right level of flavour married with a fabulous aroma. A top class oil at a reasonable price. (EW)

SANTA TEA CULTIVAR FRANTOIO
Italy

Appearance Clear and bright

Colour Green gold

Aroma Green grass

Taste Good firm green grass, up front with quite a peppery finish

Uses Drizzled over soups, casseroles, grilled vegetables, fish, mashed potato

Price £17.00 / 500ml

Comments An estate bottled oil from Tuscany and as the name suggests it's made from a single variety of Frantoio olives, hence the premium price. I like the assertiveness of this oil with its lovely green flavours. (GFF)

SANTA TEA FRUTTATO INTENSO
Italy

Appearance Clear and bright

Colour Dark gold

Aroma Bitter distinctive green leaf

Taste Fruity, green leaves, a lovely lingering taste of olive and just the right touch of pepper

Uses Drizzled over vegetables or salad leaves, soups or pasta

Price £16-£17.00 / 750ml

Comments An immediately appealing well balanced, well made oil from Tuscany. Happily use this any time and worth every penny. Treat yourself. (GFF)

BADIA A COLTIBUONO
Italy

Appearance Clear and bright

Colour Dark rich gold

Aroma Immediate green grass, good strong aroma

Taste An interesting sweet fruity flavour but too greasy on the palate. The fruit fights through in the end

Uses Salad dressings, pour over vegetables, grilled fish, soups

Price £14.25 / 500ml

Comments Disappointing oil given its lovely aroma. I wouldn't recommend it. (OM)

AVIGNONESI
Italy

Appearance Clear and bright

Colour Golden green

Aroma Light, earthy, olivey

Taste Chocolate and almond. Light subtle flavours, disappear quickly

Uses Salads, drizzled over cooked vegetables or fish, mashed potatoes, asparagus, any delicate flavoured dishes

Price £12.90 / 500ml

Comments Made from Correggiolo, Moraiolo and Leccino olives. Packed in the most remarkable black bottle. This would obviously make a stunning present. The oil is lovely and light. (EW)

DELL'UGO
Italy

Appearance	Hazy - unfiltered
Colour	Golden green
Aroma	Delicate olive, hint of banana, rich ripe - not a deep aroma but light and pleasant
Taste	Light well rounded olive taste with a mild pepper finish, not a pronounced flavour but leaves a pleasant taste
Uses	Drizzled over salads, vegetables, soups or grilled fish
Price	£12.30 / 750ml bottle
Comments	A good lighter style estate bottled Tuscan oil made from Frantoio, Moraiolo and Leccino olives. (OM)

GRAPPOLINI IL GENTILE
Italy

Appearance Clear and bright

Colour Gold

Aroma Light green leaf

Taste Sweet and lightly fruity, subtle, delicate, touch of pepper on the finish

Uses Salad leaves, steamed fish, drizzled over soup

Price £4.90 / 500ml; £6.70 / 750ml

Comments As you would expect from the name this is a light style of oil. It's very attractive, well made, well balanced, harmonious. Also a plus is the dark bottle which protects the contents from the light. From Tuscany. (FC)

GRAPPOLINI I'TOSCANO
Italy

Appearance Clear and bright

Colour Golden / green

Aroma Strong, assertive green leaf - lovely

Taste Fruity, robust green flavour, pepper finish

Uses Grilled meat, salad leaves, grilled vegetables, bruschetta, soups, casseroles

Price £9.20 / 750ml

Comments Dark bottle. A very good example of the gutsy full style Tuscan oil. (FC)

GRAPPOLINI IL GIOVINE
Italy

Appearance Clear and bright

Colour Gold

Aroma Good fresh green leaves

Taste Light, subtle

Uses Pasta, soups, salads

Price £8.00 / 750ml

Comments Another oil packed in a dark bottle to stop light, which can damage olive oil. This is a subtle harmonious oil, very well made and I liked the style a lot. From Tuscany. (FC)

PRIMOLI I SIGNORE DI TOSCANA
Italy

Appearance Clear and bright

Colour Golden

Aroma Strong fresh green grass, assertive

Taste Fruity green grass, pepper finish

Uses Drizzled on salad leaves, soups, grilled vegetables, poured over grilled meats or roasts or for bruschetta

Price £13.50 / 750ml

Comments Medium-weight oil, well made, well balanced, not too assertive. (FS)

PODERE COGNO
Italy

Appearance Thick and opaque - unfiltered

Colour Murky green

Aroma Wonderful strong, immediate green apple

Taste Lovely rich green taste with pepper finish

Uses Salads, soups, grilled fish, vegetables

Price £8.95 / 500ml

Comments This is a delicious estate bottled oil from Tuscany - not as gutsy to taste as the aroma may suggest. It has a harmonious, well balanced, attractive style. A very nice oil indeed. (HN)

RAINERI PRELA
Italy

Appearance Cloudy - unfiltered

Colour Yellow

Aroma Faint green leaf

Taste Light, subtle, not a great deal of flavour

Uses Salads, vegetables, fish or soups

Price £14.99 / 1 Litre

Comments Stunning dark green numbered bottles but the resulting oil within is disappointing at this price. (FC)

TAYLOR AND LAKE
Italy

Appearance Bright

Colour Greeny dark gold

Aroma Fresh cut green grass or hay, bitter tones

Taste Distinctive Tuscan style with a medium pepper finish. Sweet and grassy, not at all bitter

Uses Drizzled over cooked vegetables, poured over pasta, bruschetta, casseroles, grilled meat

Price £11-£12.00 / 700ml

Comments A simple stylish uncomplicated Tuscan oil. I like its lack of assertiveness and easiness on the palate. Made from Frantoio and Moraiolo olives grown in the Apennine hills. Thoroughly recommend this. (TL)

CARLUCCIO'S TUSCAN
Italy

Appearance Clear and bright

Colour Golden

Aroma Light, green, fruity

Taste Lovely, immediate, green and fruity with a hint of pepper - a light style

Uses Salads, grilled fish, raw sauces, salsa verde, pesto, pasta, vegetables

Price £7.00 / 500ml

Comments I like this, it's well balanced, not at all gutsy like some Tuscan oils. It's in the lighter style which to me is very attractive. (CO)

SANTAGATA
Italy

Appearance Clear and bright

Colour Golden

Aroma Deep and fruity

Taste Lovely rich fruit

Uses Salads, bruschetta, grilling meat and fish, pasta, soups and casseroles

Price £3.00 / 500ml; £5.35 / 1 Litre

Comments Really is a fabulous oil for the price. It's well rounded and harmonious. I thoroughly recommend this for your every-day commercial Italian brand, which you can use on everything. (FC)

LAUDEMIO SONNINO

Italy

Appearance Clear and bright

Colour Golden

Aroma Faint green grass

Taste Deep rich tone of chocolate and burnt nuts

Uses Drizzled over vegetables, casseroles, soups or pasta

Price £14.00 / 500ml

Comments A pleasant enough oil but not at this price.

LAUDEMIO BAGGIOLINO
Italy

Appearance Clear and bright

Colour Golden green

Aroma Strong green grass

Taste Quite strong pepper finish, fruity but not very distinctive

Uses Salads, grilled meat, casseroles

Price £14.00 / 500ml

Comments I don't think it is worth the money. A clear case of paying for a very fancy bottle, special pourer and a box - a whole lot of packaging dressing up a rather ordinary oil. I wouldn't buy it.

LAUDEMIO SANTEDAME
Italy

Appearance Clear and bright

Colour Golden

Aroma Sweet and fruity

Taste Sweet and fruity but mild

Uses Drizzled over fish or vegetables, soups and salads

Price £14.00 / 500ml

Comments Really rather ordinary. I've tasted better at half the price. I wouldn't buy it.

HARVEY NICHOLS EXTRA VIRGIN
Italy

Appearance	Clear and bright
Colour	Golden
Aroma	Subtle fruity olive
Taste	Sweet and fruity with a distinct taste of fresh olive
Uses	Salads, poured over cooked vegetables, fish or meat, pasta
Price	£7.45 / 500ml; £13.50 / 1 Litre
Comments	A simple straightforward and very attractive oil which is not assertive. Good for regular use and good value in this upper price range.

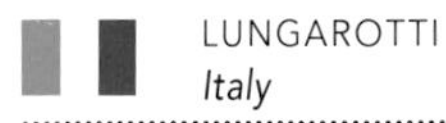

LUNGAROTTI
Italy

Appearance Clear and bright

Colour Dark olive gold

Aroma Delightful green leaf with a sharp edge

Taste Lovely and fruity, medium weight

Uses Use this raw to pour over cooked vegetables, such as boiled new potatoes, over pasta, or vegetable soups. Drizzle on salads

Price £9.00 / 500ml

Comments Well made, well balanced and an excellent example of a top quality oil from Umbria. Unlike many of the Tuscan oils you do not get any peppery finish which is a plus for me. It is also attractively packaged in a gift box making it ideal as a present. (A)

TENUTA DI SARAGANO
Italy

Appearance Clear and bright

Colour Golden

Aroma Light, subtle fruit and a hint of chocolate

Taste Light bitter chocolate. Touch of pepper at the finish

Uses Drizzled over everything

Price £13.65 - £14.25 / 750ml

Comments I like this style of oil which comes from Umbria very much indeed. It is light, well balanced with the warm tones of this interesting chocolatey taste. Estate bottled and made from Frantoio, Moraiolo and Leccino olives. Numbered and dated bottles. (OM)

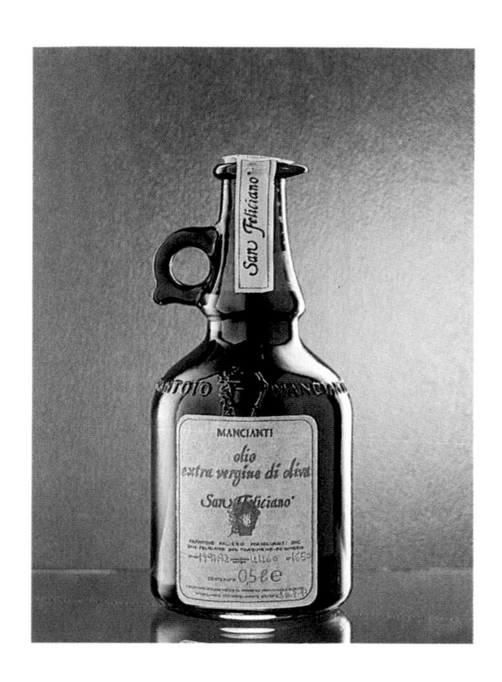

MANCIANTI SAN FELICIANO
Italy

Appearance	Clear and bright
Colour	Golden
Aroma	Light subtle green grass
Taste	Fruity, hint of pepper finish
Uses	Salads, grilled vegetables, bruschetta, soups, casseroles
Price	£9.00 / 500ml
Comments	Lovely rich fruity oil, just the right amount of pepper for me. From Umbria. (GFF)

FATTORIE UMBRE CAMPOMAGGIORE
Italy

Appearance Clear and bright

Colour Rich green

Aroma Bitter, full green apple and leaf

Taste Avocado

Uses Drizzled on cooked vegetables, sliced tomatoes - would be marvellous for tomato and mozzarella salad

Price £7.50 / 500ml

Comments A truly amazing taste like eating a ripe avocado. Fabulous oil from Umbria which I strongly recommend. (GFF)

TEGA CARDINALE
Italy

Appearance Clear and bright

Colour Greenish gold

Aroma Very faint green olive

Taste Chocolate and bitter almonds with a slight touch of pepper

Uses Salads, drizzled over cooked vegetables or meat, soups or casseroles

Price £9.30 / 500ml

Comments Lovely deep woody taste while still being light in style. This is a lovely oil from Umbria made from a blend of Moraiolo, Frantoio and Leccino olives. (GFF)

MANCIANTI AFFIORATO
Italy

Appearance Clear and bright

Colour Golden

Aroma Strong, sharp, green leaves, lovely full verdant tones

Taste A little greasy, not much taste but sweet and fruit

Uses Salad dressings, grilling fish or meat

Price £12 - £13.00 / 500ml

Comments A nice enough oil but not sensational.
Comes from Umbria. (GFF)

SASSO
Italy

Appearance Clear and bright

Colour Yellow

Aroma Good strong green grass

Taste Light, pleasant taste of fruit, very mild with a peppery finish

Uses Salad dressings, grilling fish or meat

Price £3.00 / 500ml; £5.50 / 1 Litre tin

Comments A simple, straightforward, uncomplicated everyday commercial extra virgin with a subtle taste. Great for those people who do not like strong assertive olive oils. Also comes in a litre tin which is ideal for best protection against sunlight. All round good value oil and it is one of my favourite commercial blends of Italian oil in this price range. (A)

ROCCHI ORO
Italy

Appearance Clear and bright

Colour Golden

Aroma Fruity

Taste Fruity and light with a pleasant hint of bitter green olive

Uses Salads, for grilling or drizzled over vegetables

Price £2.50 / 500ml

Comments Not at all heavy or assertive, a very nice style of Tuscan oil. Well made and harmonious. An excellent example of what a good commercial oil should be like. Bottled in Lucca at a price to match supermarket brands making it exceptionally good value. Recommended. (A)

BERTOLLI
Italy

Appearance Clear and bright

Colour Green

Aroma Ripely fruity

Taste A little fruity taste but fades away fast, a hint of pepper

Uses Roasting or grilling

Price £3.95 / 1 Litre

Comments Not very exciting, fruitier than some but you can't complain at this price.

FILIPPO BERIO SPECIAL SELECTION
Italy

Appearance Bright and clear

Colour Golden

Aroma Immediate green grass, good assertive aroma

Taste Slightly greasy start, peppery finish rather flat not much of anything

Uses Salad dressings, roasting, grilling

Price £5.00 / 750ml

Comments The Berio oils have huge sales but I personally think they're disappointing. I find them greasy, with very little flavour and I know of much better Italian oils in this price range.

FILIPPO BERIO
Italy

Appearance Bright

Colour Golden greeny yellow

Aroma Pleasant, bitter with a hint of pepper

Taste Flat and greasy, very little flavour but a mild pepper finish

Uses Salad dressings, grilling meat or fish, marinades

Price £2.52 / 500ml

Comments To my mind there are many better Italian oils on the market at this price and even though it's a brand leader, I couldn't recommend it. It has very little taste and is, I think, considering its reputation, most disappointing.

CALLISTO FRANCESCONI
Italy

Appearance Clear and bright

Colour Greenish gold

Aroma Good sharp green grass

Taste Immediate grassy flavour not too strong, hint of pepper at the finish

Uses Salads, pouring over cooked vegetables or casseroles, grilling meat or fish, marinading, pasta dishes

Price £3.75 / 500ml; £6.49 / 1 Litre

Comments This is a well balanced, harmonious and tasty oil bottled in Lucca. Simple and basic for everyday use. It is one of my top choices for a commercial brand Italian oil in this price range. Recommended. (PW)

PACELLI
Italy

Appearance Clear and bright

Colour Golden

Aroma Faint green grass

Taste Light, subtle taste with a hint of green olive and some pepper on the finish

Uses Salad dressings, grilling meat or fish

Price £5.00 / 1 Litre

Comments A simple oil. Well made and well balanced. Perfectly good for ordinary everyday use if you like the lighter style oils. (N)

DANTE
Italy

Appearance Clear and bright

Colour Golden

Aroma Good strong bitter green grass

Taste Rather tasteless, just a hint of pepper finish

Uses Salad dressings, grilling or roasting meat, grilling fish

Price £2.55 / 500ml

Comments This olive oil has a very high profile, supported by lots of advertising and is stocked by many supermarkets. I was surprised therefore to find it such a disappointment; with its lovely aroma it promised much but with so little taste I can't recommend it. There are other much better Italian oils in this price range.

ST MICHAEL EXTRA VIRGIN
Italy

Appearance Clear and bright

Colour Golden

Aroma Strong green grass

Taste Strong peppery finish, green grass. A little greasy but not enough to mar the good flavour

Uses Salad dressings, roasts, grills

Price £2.39 / 500ml

Comments Well made commercial blend. Pretty good for the price. Useful oil for everyday use.

SAINSBURY'S EXTRA VIRGIN DI CANINO
Italy

Appearance Clear and bright

Colour Greenish gold

Aroma Faint hint of green grass

Taste Light buttery flavour. Hint of pepper on the finish

Uses Salad dressings, marinades

Price £2.29 / 250ml

Comments This is, according to the label, a Tuscan single estate oil. It also says that it is a rich dark green oil with an intense fruity flavour and aroma. Well I must be colour blind and have lost my sense of taste and smell! There's nothing wrong with the oil - it's okay for the price but it certainly isn't how it's described. I prefer their cheaper ordinary extra virgin.

SAINSBURY'S EXTRA VIRGIN
Italy

Appearance Clear and bright

Colour Golden

Aroma Sweet, fruity, rich

Taste Green, light, mildly fruity. Flavour disappears quite quickly

Uses Salad dressings, grilling, roasting

Price £2.12 / 500ml; £3.85 / 1 Litre

Comments Pretty good oil for the price - it's a pity it doesn't have a little more flavour after the lovely aroma but a good all purpose cheap everyday oil.

WAITROSE EXTRA VIRGIN
Italy

Appearance	Clear and bright
Colour	Golden with a hint of green
Aroma	Green and grassy
Taste	Light peppery aftertaste but almost no discernible taste
Uses	Roasting, grilling and salad dressing
Price	£2.79 / 500ml
Comments	I couldn't recommend this on the grounds there is really no flavour of any kind.

SAFEWAY EXTRA VIRGIN
Italy

Appearance	Clear and bright
Colour	Golden
Aroma	Pleasant fresh green grass but not very strong
Taste	Not too robust at all, just leaves a fruity hint of green grass
Uses	Grilling meat or fish, making salad dressings
Price	£2.49 / 500m
Comments	Good basic supermarket brand. Excellent for everyday use where you don't want an assertive flavour.

MERIDIAN
Portugal

Appearance Hazy

Colour Deep yellow gold

Aroma Sharp green leaves but not very strong. Not altogether pleasant

Taste Greasy and flat, leaves a rather unpleasant taste of earth in the mouth

Uses I wouldn't

Price £3.45 / 500ml

Comments I really did not like this oil. I thought at first it was rancid so I opened and tasted a second bottle and it tasted just the same. This would appear just to be the style of Portuguese oil and I have to confess it's not to my taste at all. It's a shame since many people would be drawn to it because it's organic. But there are other organic oils available.

LA ROSA
Portugal

Appearance Clear and bright

Colour Green gold

Aroma Earthy, woody strong pungent smell of warm compost. Not altogether pleasant to my senses

Taste Bitter, earthy, flat

Uses Salad dressings or grilling meat and fish

Price £4.50-£5.00 / 375ml; £9.50-£10.00 / 750ml

Comments I have tasted two Portuguese extra virgin oils and I haven't liked either. Maybe it's the style, to me their flavours seem unpleasant. (B)

NUNEZ DE PRADO
Spain

Appearance	Hazy - unfiltered
Colour	Golden
Aroma	Rich ripe luscious melon mixed with green grass. Wonderfully fruity floral aroma
Taste	Sweet light tropical fruit, peppery aftertaste. Gorgeous, gorgeous, gorgeous
Uses	I pour this on everything, salads, boiled potatoes, grilled fish, steamed vegetables
Price	£8.00 / 500ml
Comments	This is a fabulous example of an estate bottled Spanish oil and it's organic. It is so refreshing. There is nothing heavy or cloying about this oil. It sits lightly and fruitily on the palate. My only complaint is the red sealing wax on the cork which goes everywhere but I really do recommend it most highly. (B)

ORO MAGINA
Spain

Appearance Clear and bright

Colour Deep gold

Aroma Gorgeous, heady perfume of ripe melons

Taste Sweet, ripe fruit. It's light and delicious

Uses On everything! But don't cook with it. Try it in mashed potato or poured over cooked dishes

Price £4.95 - £5.50 / 500ml

Comments This is a wonderful estate bottled oil from Jaen made from Picual olives. Excellent quality, beautifully balanced. I could use this all the time and not tire of it. Fabulous value for money, it must be one of the cheapest estate bottled oils around. Worth buying at twice the price. (AM)

SIURANA
Spain

Appearance Bright and clear

Colour Deep golden

Aroma Sharp, sweet freshly mown green grass

Taste Lovely light sweet sunny taste

Uses Pour over cooked vegetables, drizzle over salads

Price £6.40 / 1 Litre

Comments I am very taken with this style of oil from Catalonia and I'd describe it as fun and jolly and I'd use this often. Excellent, well made, well balanced and exceptional value for money. Very different style to the Andalucian oils. Thoroughly recommend. (B)

LERIDA
Spain

Appearance Clear and bright

Colour Golden

Aroma Very faint fresh olives

Taste Chocolatey burnt nuts, just a hint of pepper

Uses Drizzled over baked potatoes, or grilled vegetables or rich casseroles

Price £7.90 / 500ml; £12.85 / 1 Litre

Comments This is a most unusual and delightful tasting oil, not quite like anything else. Made from Arbequina olives grown in Catalonia. Estate bottled.I would describe it as darkly moody. If you're a chocoholic you'll love this. Good value too. (OM)

L'ESTORNELL
Spain

Appearance Clear and bright

Colour Deep gold

Aroma Light subtle hint of almonds

Taste Sweet with a slight peppery finish

Uses Salads, poured over cooked vegetables or fish, soups or any hot dishes

Price £8.50 / 750ml

Comments This is an enjoyable, simple estate bottled Spanish oil made from Arbequina olives. Coming from Lerida in Catalonia it is a different style to the rich, heady tropical aroma of oils from the south. It is certified as organic which will make it popular with many people. (OM)

CARBONELL
Spain

Appearance Clear

Colour Rich golden

Aroma Fresh, sweet, luscious ripe tropical fruit such as melon

Taste Sweet taste of tropical fruit. Well rounded not overpowering

Uses On salads, grilled vegetables, soups, casseroles. Lovely on baked potatoes

Price £2.49 / 500ml; £3.99 / 750ml

Comments This is a wonderful oil and has long been a favourite with me. I never tire of it for using everyday. It is exceptional value for money and I strongly recommend it. (C)

BORGES
Spain

Appearance Clear and bright

Colour Golden

Aroma Sharp, fresh apple skins and a faint hint of pear drops

Taste Light, simple taste of green olives

Uses Salad dressings, grilling, roasting

Price £2.20 - £2.50 / 500ml

Comments This is a basic everyday oil, lighter in style than some Spanish oils and very good value. Equal in price to most supermarket own-brands. (U)

SAFEWAY EXTRA VIRGIN
Spain

Appearance Clear and bright

Colour Golden yellow

Aroma Good sweet tropical fruit, quite reminiscent of lychees

Taste Bit greasy at first but the flavour comes through gently and leaves a pleasant taste, slightly almondy

Uses Salad dressings, marinades, for grilling or roasting meat or fish

Price £2.49 / 500ml

Comments Not the very best example of a Spanish oil but it's not bad by any means given the price.

WAITROSE EXTRA VIRGIN
Spain

Appearance Clear and bright

Colour Yellow gold

Aroma Rich ripe melon - gorgeous aroma

Taste A little bit greasy but it finishes to typical Spanish style of a sweet and fruity taste

Uses Salad dressings and on cooked vegetables

Price £2.49 / 500ml

Comments A very typical distinctive Spanish oil. Good for everyday use at the price.

SAFEWAY EXTRA VIRGIN

EEC

Appearance Clear and bright

Colour Light green

Aroma Light, green grass, sweet

Taste Slightly greasy but not unpleasant. Light, not much flavour

Uses Salad dressings, grilling or roasting meat

Price £2.95 / 500ml

Comments Not a bad oil for the price though there is not really much of a flavour. There are better at this price

ASDA EXTRA VIRGIN
EEC

Appearance Clear and bright

Colour Golden

Aroma Faint but earthy

Taste Pungent, green leafy

Uses Salad dressings, grilling, roasts

Price £2.12 / 500ml

Comments Good for everyday use at the price. Pretty good example of commercial EV and as the label suggests it's a blend of oil from several countries. They are to be commended for being honest about that.

TESCO EXTRA VIRGIN
EEC

Appearance Clear

Colour Rich dark gold

Aroma Fruity with a hint of tropical fruit or ripe bananas. Aroma leads me to suspect a large proportion of Spanish oil

Taste A sweet fruity start with a peppery finish

Uses Salad dressings, grilling or roasting meat, fish or vegetables

Price £2.12 / 500ml

Comments I would think that this is a blend of Spanish and Italian oils since it is described as produce of EEC. This is a pretty good oil - excellent value at the price.

OLIO SANTO

USA

Appearance Clear

Colour Golden with greenish tinge

Aroma Sweet, nutty

Taste Reminiscent of pear drops. Very light, slightly greasy sensation

Uses Salads, pouring over cooked vegetables or fish

Price £9.95 / 500ml

Comments As the only US oil in my tastings it's difficult to judge the style of this oil against anything but it's nothing extraordinary and I can think of other oils I'd prefer at this price. (OM)

LATE ENTRIES

There are new extra virgin olive oils being introduced to the shops almost every week and so it was inevitable that there would be some which came to my attention as this book was going to print or some where tasting samples came in too late to be photographed. In order to be as up to date and comprehensive as possible I have included some with brief comments. If there are oils you use and like which are not in the book please write to me via my publisher, Grub Street and let me know.

Athena: a Greek oil from Kalamata. Another example of a good every-day oil. Now available from some supermarkets. (U)

Land of Canaan: an olive oil from Israel, so rather unique. Well made, and harmonious. It is a Kosher oil from Hallgarten. Tel 0582 22538.

Mennucci: a reasonable commercial blend of Italian oil at the cheaper end of the price range.

Napolina: disappointing, hardly any flavour. Another bland Italian commercial blend.

Masseria: lovely silky smooth oil from Apulia, with deep fruity tones and a peppery finish that sneaks up on you. Excellent price for such quality. (D)

Roi: so light it bounces round the mouth, such a delicate sweet flavour and typical of these wonderful Ligurian oils. Excellent. (D)

SHOPS AND SHIPPERS

Every supermarket now carries a range of low priced olive oils but if you want real choice and variety then delicatessens and specialist food shops are the places to explore. In these you are most likely to find some of the best olive oils and certainly single estate oils at the premium price end of the market and if you're lucky, the owners may be knowledgeable enough to advise you. If they don't know anything about the oils they have in stock why not suggest they organize an olive oil tasting.

Here are some of the shops I consider to have good ranges of oils. Also a list of some importers and distributors, who may prove useful if you want to know which outlets local to you are stocking their oils or indeed if you own a shop and wish to buy wholesale. The names have been coded and appear in the tasting notes section so that you can identify the source of many of the oils photographed. The more commonly available oils are not coded.

These lists are by no means exhaustive, so please do write and tell me if there is a shop you know of which should be included in the next edition.

LONDON

Harvey Nichols Foodmarket
Fifth Floor
Knightsbridge, London SW1
☎ 071 581 7594
Surely the most spectacular range of olive oils stocked anywhere, in one of the most gorgeous food emporiums. At last count they had over 60 different extra virgin olive oils. Mark Lewis, the buyer is immensely knowledgeable and helpful. They have a number of exclusive oils including their own brand and they run regular tastings and tutorials.

Vivian's
2 Worple Way
Richmond, Surrey
☎ 081 940 3600
A gem of a shop run by the eponymous Vivian, a man who knows his oils and his mind. He deserves a medal for selling some of the best extra virgin oils loose. You supply the bottle and he'll fill it. It's a brilliant idea and one that more shops could usefully employ.

The Realfood Store
14 Clifton Road
Little Venice, London W9 1SS
☎ 071 266 1162
Another shop with an innovative sales technique. They supply tasting samples of any of their single estate and regional first cold pressed extra virgin oils,in small bottles for £1.00.

Fratelli Camisa
53 Charlotte Street
London W1
☎ 071 255 1240
They always stock a wide range of olive oils, especially Italian. There are regular tastings and they now have a mail order catalogue so you can order by post. They also wholesale a number of their exclusive lines.

Villandry
89 Marylebone High Street
London W1
☎ 071 487 3816
Many, many wonderful exclusive EV olive oils.

Carluccio's
28 Neal Street
London WC2
☎ 071 240 1487
The shop is run by Priscilla Carluccio and she is a great discoverer of Italian EVs. There are at least three own-brand oils from different regions of Italy and she is constantly seeking new ones.

Selfridges
400 Oxford Street
London W1
☎ 071 629 1234
Always an excellent food hall but now with a big olive oil section nurtured by the caring and informed manager, Tony Greenwood.

The Conran Shop
Michelin House
81 Fulham Road
London SW3
☎ 071 589 7401

Tom's
226 Westbourne Grove
London W11
☎ 071 221 8818

Mortimer and Bennett
33 Turnham Green Terrace
London W4
☎ 081 995 4145

Le Pont de la Tour Oil and Spice Shop
Butlers Wharf Building
Shad Thames, London SE1
☎ 071 403 3434

Hamish Johnston
48 Northcote Road
London SW11
☎ 071 738 0741

Clarke's
122 Kensington Church Street
London W8
☎ 071 229 2190

LANCASHIRE
The Ramsbottom Victuallers
16-18 Market Place
Ramsbottom, Bury
☎ 0706 825070
Chris Johnson's shop is certainly compact but a more excitingly stocked outlet for its size you'd have difficulty finding. He has an excellent range of EV olive oils.

CUMBRIA
J&J Graham
Market Square
Penrith

SUFFOLK
Adnam's
The Kitchen Store
Victoria Street
Southwold

NORFOLK
Humble Pie
Burnham Market

OXFORDSHIRE
Jordans
8 Upper Street
Thame

Fasta Pasta
121 The Covered Market
Oxford

BIRMINGHAM
Rackhams
Corporation Street

LEICESTER
David North Ltd
289 Station Road
Rothley

DEVON
Crebers of Tavistock
48 Brook Street
Tavistock

SUSSEX
Comestibles
Church Hill
Midhurst

SCOTLAND
Valvona and Crolla
19 Elm Row
Edinburgh

Fratelli Sarti
133 Wellington Street
Glasgow

EIRE
Ballymaloe Shop
Shanagarry, Co Cork

WALES
Howells
14-18 St Mary's Street
Cardiff

DISTRIBUTORS
The letters after each entry indicate the origin of the oils distributed. F - France, Gk - Greece, It - Italy, P - Portugal, Sard - Sardinia, Sp - Spain, US - United States

A Alivini ☎ 081 880 2525 (It)
AM Amber Foods ☎ 0208 815309 (Sp)
B Brindisa ☎ 071 403 0282 (Sp, P)
BN Bevelynn ☎ 0992 641441 (Gk)
C Carbonell ☎ 081 891 5015 (Sp)
CO Carluccio's ☎ 071 240 1487 (It)
D Danmar ☎ 081 844 1494 (It)
EC Eurochoice ☎ 081 653 9422 (Sard)
EW Eurowines ☎ 081 994 7658 (It)
F The Fresh Olive Company of Provence ☎ 081 838 1912 (F)
FC Fratelli Camisa ☎ 071 255 1240 (It)
FS F&S Group ☎ 081 205 3669 (It)
FTC The Frenchland Trading Company ☎ 0903 879471 (F)
GFF Guidetti Fine Foods - *Enormous range of Italian EV olive oils* ☎ 081 460 3727
GS George Skoulikas ☎ 081 452 8465 (Gk)
HN Harvey Nichols ☎ 071 581 7594
I Italbrokers ☎ 071 627 0030 (It)
JP John and Pascalis ☎ 081 452 0707 (Gk)
K Kastouris Brothers ☎ 071 607 2730 (Gk)
LL Leathams Larder ☎ 071 252 7838 (F,Gk,It)
N L Noel and Sons ☎ 0254 301324 (It)
O Odysea Ltd ☎ 071 256 8668 (Gk)
OM The Oil Merchant (wholesale and mail order). *The broadest range of olive oils from the man who really put olive oil on the map in this country.* ☎ 081 740 1335 (F,Sp,It,Gk,US)
PW Petty & Wood ☎ 0264 345500 (F,It)
TL Taylor and Lake ☎ 0608 683366 (F,It)
U Unimerchants ☎ 0353 661999 (Gk, It, Sp)
W Winecellers ☎ 081 871 397 (It)

INDEX TO THE OILS